CW00485036

# WordPress

## Security tips

## How to outsmart hackers

# INTRODUCTION

## Why this book?

In recent years, WordPress has become the most widely used Content Management System (or CMS) in the world. In fact, according to the latest W3Techs barometer, WordPress now powers more than 30% of the websites in the world!

WordPress has many advantages, it is an open source, free, and the community and publishers have developed plugins to do everything: social network integration, forum, e-commerce website, online payment, photo gallery, media player, SEO, calendar, survey, hotel booking... Its use has been very democratized by the many books and websites that are flourishing and that explain step by step how to install it, how to start and build your first site with WordPress, how to customize your WordPress site, build an e-commerce site, etc... It is even possible to use themes (free or paid) to get, in a few clicks, a professional website worthy of a web agency.

In short, it now seems impossible not to consider WordPress when it comes to building websites, especially if you are not a professional in the field of web development, because you will be able to put a beautiful website online without having to write a single line of code.

However, there is another side to this medal, rarely mentioned, which motivated the writing of this book: the computer security of WordPress sites. Indeed, like all computer software, WordPress has security flaws. Automatically activated security updates are not enough to solve the problem because they do not cover the many vulnerabilities in the plugins and themes that are the responsibility of their respective publishers. Besides, some security holes are not technical but "human", such as using the WordPress admin account with a trivial or easily crackable password such as "admin123".

Since WordPress is widely deployed around the world and often used by amateur webmasters, websites that use it are logically very exposed to attacks by hackers from around the world. These hackers regularly succeed in doing a lot of damage because the rules of the trade and good practices in terms of computer security have not been applied. **Among the CMS sites hacked last year, 90% were WordPress sites!**

Building websites with WordPress for the past 10 years, I have often noticed the lack of knowledge of security issues when "disinfecting" and securing hacked websites. This legitimate lack of knowledge motivated me to write this book. My goal is to democratize good security practices and to propose concrete actions, easy to realize on your WordPress site to significantly strengthen your website against attacks.

Even if your site is not that of a multinational company, it is likely to be attacked. Actually, hackers launch automated scripts (or software robots) that exploit the flaws in WordPress or certain plugins for:

- Adding inappropriate content to the website (e.g. links to illegal websites)
- Adding spam comments
- Destroying the content of the website
- Crashing the website or slow down the website (this is called a Denial of Service attack or DoS attack)
- Extract information (example: retrieve the list of emails of people who have left a comment)
- Injecting invisible scripts that will infect the browsers of users visiting your site

The tools and operating methods to carry out these malicious operations are unfortunately easily accessible on the Internet and it is no longer necessary to be an experienced computer specialist to carry out hacking actions. Someone can screw up your website, for fun, to harm you, or steal information about your users (I will come back to this later).

## Who is this book for?

You don't need to be a computer wizard to read and apply the principles and actions explained in this book. Like the many books on WordPress, I wanted this one to be understandable and applicable by the greatest number of people. My goal is that as many sites as possible should be more secure! The technical aspects are thus well explained and computer readers will find the opportunity to revise, and maybe sometimes, another point of view on a subject already known.

## What will you find in this book?

I wanted this book to be practical and easy to use, because if you read it without implementing the measures set out in it, it will be useless! That is why it has been conceived as a succession of short chapters explaining why and how to protect yourself from a specific loophole. I have placed particular importance on explaining the concepts used by the attackers so that you understand why I am recommending one measure or the other.

Fixing some vulnerabilities requires plugins, while for others it is not necessary. If necessary, I will always propose to you free plugins (or freemium) to protect your site.

You will see that I have detailed step-by-step the settings or configuration changes to be made. To help you, when necessary, I've integrated screenshots, but WordPress and its plugins evolve quickly and even if the menus and sections remain the same, you may find discrepancies between the screenshots in this book and what you'll see at home.

## I send you an email with all the settings to copy/paste

To make things easier, I propose to send you by email all the settings and links to the plugins that are referenced in this book. All you have to do is copy/paste the settings into your WordPress instance and you will avoid a tedious re-typing of the lines in the book. To do this, send an email to **wpbonus@gmx.com** with "BONUS WP" in the subject line, and I will send you a message with all the information.

## By the way, who am I?

My name is Thomas Person and I am a computer engineer with 18 years of experience. I am currently working on IT security issues in a very large company. In parallel to my work, I have been using WordPress since 2010, first to create personal sites, then for family and friends, and now for "real" clients with whom I sign contracts.

Over the years, I have tested and implemented many security solutions on my WordPress sites, and seeing that there was no French book dealing with this issue, I decided to write one.

Since its origins as a blog engine, WordPress has evolved a lot to become a full-fledged CMS (Content Management System), and taking into account IT security issues is now a must considered aspect.

## First of all, a few reminders on IT security

Cybersecurity is about protecting computers, servers, mobile devices, electronic systems, networks and data from malicious attacks. These attacks can be of several types:

1 - Shutting down a website

A denial-of-service attack (DoS attack) is a cyber-attack in which the perpetrator seeks to make a machine or network resource unavailable to its intended users by temporarily or indefinitely disrupting services of a host connected to the Internet. Denial of service is typically accomplished by flooding the targeted machine or resource with superfluous requests in an attempt to overload systems and prevent some or all legitimate requests from being fulfilled.

In a distributed denial-of-service attack (DDoS attack), the incoming traffic flooding the victim originates from many different sources. This effectively makes it impossible to stop the attack simply by blocking a single source.

A DoS or DDoS attack is analogous to a group of people crowding the entry door of a shop, making it hard for legitimate customers to enter, thus disrupting trade.

In 2012, not one, not two, but a whopping six U.S. banks were targeted by a string of DDoS attacks. The victims were no small-town banks either: They included Bank of America, JP Morgan Chase, U.S. Bancorp, Citigroup and PNC Bank. The attack was carried out by hundreds of hijacked servers, which each created peak floods of more than 60 gigabits of traffic per second.The damage in terms of brand image and loss of earnings has been considerable.

2 - Stealing users' personal information

In September 2018, the airline company British Airways had the financial data of 380,000 customers stolen. For each of them, the following information was stolen: name, postal address, email and above all, their credit card data, i.e. number, expiry date and the secure three-digit code. This data is generally resold by hackers on the dark web.

3 - Steal confidential information of companies

In October 2016, hackers stole the personal data of 57 million customers and drivers from Uber through a massive breach.

4 - Altering the content of a website

In January 2020, the website of a US government agency was hacked. The information on the site has been replaced by pro-Iranian content showing President Trump being punched in the jaw.

5 - Identity theft

In April 2016, hackers linked to the Syrian regime took control of the Twitter account of the French newspaper "Le Monde".
Many other attacks take place on a regular basis, but the victims do not always make them public, as this can affect the company's share price or image with the public, or they simply never knew that they had been hacked.

# Chapter 1: Why and How to Protect Your WordPress website

## Number 1: An essential measure: The use of a "strong" password

Hackers use software robots that scan as many websites as possible and for each of them search if a /wp-admin page exists.

If the /wp-admin page exists, then they try to authenticate to the WordPress site using the admin login and a list of frequently used passwords, gathered in a file called a dictionary. This type of attack is called a "dictionary attack" and consists of **testing a series of potential passwords, one after the other**, hoping that the password used is in the dictionary.

The dictionary attack is thus based on certain common dictionaries, such as dictionaries of first names, of surnames of a country or culture, of animal names, of frequently used passwords (welcome, football, password, iloveyou, 123456, azerty, abc123...). Hackers with few technical computer resources thus easily discover passwords constructed with existing words.

Based on more than 5 million passwords leaked during the year, SplashData has published a list of the most used passwords: 123456, Password, 12345678, qwerty, 12345, 123456789, football, iloveyou, admin, login, abc123, starwars, 123123, dragon, password...

You can see why dictionary attacks work well!

To have a strong password, you need a password of at least 12 characters, containing 3 of the 4 types of characters: uppercase, lowercase, number, special character (for example: @ ? ! % = ] + ...) The Cybersecurity and Infrastructure Security Agency (CISA) even recommends users:

- to use a unique password for each service. In particular, the use of the same password between your professional and personal email is imperatively prohibited

- to choose a password that is not related to you, a password composed of a company name, a date of birth, the first name of your children, etc.)

Under these conditions, your WordPress password will not be the same as the password you use, for example, on a poorly secured forum (where passwords are easily accessible), or to access your email. In 2013, when 3 billion Yahoo accounts were hacked, if you had a Yahoo account and used the same password for all your accounts, hackers would have had access to your Amazon, Apple, Google, online banking, PayPal, tax, social security, mobile phone, DropBox... and of course your WordPress sites.

If your payment card was registered on some of these sites, you can imagine the damage!

If Yahoo, a multinational company employing the best engineers, has been hacked, your WordPress site is of course hackable. In that case, if your WordPress password was hacked (without you even realizing it), at least the hacker wouldn't have "open bar" access to all your other accounts!

The interest of having a different password for each service being proven, the question arises as to how to remember all your passwords.

To this question, Marc Goodman, cyber-security expert at Interpol, then at the FBI, author of the book "Crimes of the Future", advises to use a password manager (or digital safe), such as: 1 Password (paying), LastPass (paying), Dashlane (paying), BitWarden (free)...

The principle of these solutions is relatively simple, the software at the first startup asks for a general password that will allow the unlocking of the manager (that you should never forget). You will then be able to manually enter your passwords, identifiers, notes, credit card in a totally secure way. These managers propose a plugin to be installed on your browser that will fill in the login and password fields of the sites for you. Thus, you will be able to use a different password for each site and don't need to remember it, you will only need to remember your manager's password, which will of course have to be very strong. These managers even offer you a strong password generator, which you can use to change the passwords of your sites.

The most common WordPress hacking attempts use stolen passwords (for example, those stolen from Yahoo in 2013), so **it is absolutely necessary to use strong passwords that are unique to your website**. Also think about your other passwords, those used for your FTP accounts, the database, access to your host and of course the one for your email address.

## Number 2: Blocking "brute force attacks"

Maybe you use a suitcase closed with a secret code when you go on holiday. This lock only unlocks when a certain code is entered using small knobs. For a thief who would get his hands on your luggage, the only way to open your suitcase (without special tools) would be to try, one by one, all the possible combinations. If the lock has three knobs with numbers from 0 to 9, that makes 1,000 possible codes, at a rate of 2 seconds for each attempt at a code, the thief would open your suitcase in a maximum of half an hour!

This method of finding a code is called brute force or brute force attack. A hacker can ask a computer to take an account and potentially test thousands of passwords per second. This makes short passwords without special characters particularly vulnerable. For example, a software robot performing this type of attack could be configured to try all possible passwords from 3 to 8 characters containing only letters, hoping to find yours.

This kind of attack can be identified easily by the number of unsuccessful connection attempts made from the same IP address (i.e. the unique Internet identification number of the PC or server used by the hacker).

To protect yourself from this type of attack, simply install the iThemes Security plugin (formerly Better WP Security). iThemes Security is a real Swiss army knife for the protection of your WordPress site, as of its freemium version! Among its many features such as 404 error detection, absent mode, blacklist, file change detection etc... is of course protection against brute force attacks.

Once you have installed and activated the plugin, you will need to go to "Local brute force protection". A pop-up window will then open to allow you to set up this feature:

- The maximum number of attempts per host (it is the IP address that is targeted)

- The maximum number of attempts per user (it is the user's name that is targeted)

- The lock-up period before you can try the connection again.

About Lockouts

Your lockout settings can be configured in Global Settings.
Your current settings are configured as follows:

- **Permanently ban:** yes
- **Number of lockouts before permanent ban:** 3
- **How long lockouts will be remembered for ban:** 7
- **Host lockout message:** erreur
- **User lockout message:** Vous avez été bloqué·e suite à un trop grand nombre de tentatives de connexion infructueuses.
- **Is this computer white-listed:** yes

Max Login Attempts Per Host | 5 | Attempts

*The number of login attempts a user has before their host or computer is locked out of the system. Set to 0 to record bad login attempts without locking out the host.*

Max Login Attempts Per User | 10 | Attempts

*The number of login attempts a user has before their username is locked out of the system. Note that this is different from hosts in case an attacker is using multiple computers. In addition, if they are using your login name you could be locked out yourself. Set to 0 to log bad login attempts per user without ever locking the user out (this is not recommended).*

Minutes to Remember Bad Login (check period) | 5 | Minutes

*The number of minutes in which bad logins should be remembered.*

Automatically ban "admin" user | ☐ Immediately ban a host that attempts to login using the "admin" username.

This plugin will block IPs and users making a high number of attempts to access your back office. This way, if a robot tries to enter your site, the plugin will block access to it for a certain period of time (configurable in the plugin).

Once you have installed the plugin, you will be able to set the number of tries you want before blocking and the duration of the blocking of the IP concerned. For example, you could choose to allow 10 consecutive unsuccessful authentication attempts and beyond that, block the relevant IP for 5 minutes. With this simple setting, a "brute force" type attack becomes impossible because it takes too long to drive.

On the other hand, if you have a password that is long enough (minimum 12 characters) and complex enough (mix of upper and lower case letters, numbers, special characters), your password will be said to be "strong", i.e. resistant to a "brute force" attack, because the number of possible combinations will be very high and a "brute force" attack would take too long.

## Number 3: Never use the WordPress "admin" account

If you use an account other than the admin account, hackers can always try to crack the admin account password, they will never succeed since the admin account will not exist.

If you are using the admin account, it is urgent to create another admin account and to delete the admin account, here is how to do :

Creation of the new Administrator account:
- On the Users menu, click Add
- Create a user with an identifier that only you will know
- Choose the Administrator role
- To finish, click on "Add a user".

Deleting the Administrator account "admin":

- On the Users menu, click All Users
- Position the mouse cursor on the admin account, the submenu "Edit Delete Display" appears.
- Click on Delete
- Select "Assign content to" and choose your new Administrator account from the drop-down list.
- Finally, click on "Confirm this action".

At the end of this action, there will no longer be an "admin" user account on your WordPress, any attack attempt using this account will necessarily fail.

You can further secure your site, via the iThemes Security plugin (which we saw earlier to block brute force attacks) so that it immediately bans IPs from which connection attempts are made with the username "admin". To do so, you just have to check the "Automatic ban admin user" box on the screenshot above.

## Number 4: Limit the number of administrator accounts on your site and force users to choose strong passwords

If you have several people managing the WordPress site, try to give only the necessary rights to each user. For example, if some contributors are only responsible for publishing articles, they should be given the role of Author and not Administrator. The fewer administrators on your website, the better it is for its security.

If your site uses WooCommerce, Internet users will have to create their own account to access their customer account and consult their orders, invoices... In this case, you will have to study a solution to avoid that these people use non-recommended usernames or weak passwords, even if they only have a "subscriber role" which has few rights on WordPress. The iTheme Security plugin will allow you to force users to use strong passwords.

To do so, in the "Password Requirements" section, you can choose the minimum profile for which a strong password will be required. Higher profiles will of course automatically be affected.

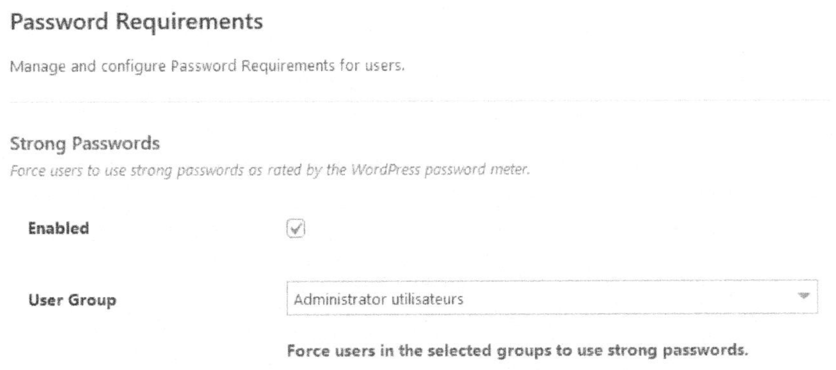

Password Requirements

Manage and configure Password Requirements for users.

Strong Passwords

*Force users to use strong passwords as rated by the WordPress password meter.*

**Enabled** ☑

**User Group** Administrator utilisateurs

**Force users in the selected groups to use strong passwords.**

## Number 5: Changing the WordPress back office access URLs (wp-admin and wp-login)

The back office corresponds to the screens you access to manage your content (create articles, pages, import images...), it is the WordPress administration panel, as opposed to the front office which corresponds to the screens accessed by non authenticated users, to read your content.

By default, the back office access url for all WordPress instances is: *http://your_domaine/wp-admin*.
This url is known to all WordPress users, and therefore to hackers who may try to connect to your back office.

By modifying it, the url will then be known only to you and it will protect the front door of your back office.

The easiest way to do this is to use the "Hide My WP" plugin, the freemium version is enough.

Once you have installed and activated the plugin, in the plugin administration panel, you will find the option "Hide wp-admin", this is one of the first options of the plugin. Once you have activated it, you will need to choose your own back office access url, which you can enter in the "Custom admin URL" field. In this field, you just have to indicate what you have chosen to replace "wp-admin", for example, if you want to access the back office of your site via the url *http://your_domaine/admin-console*, you will have to enter "-admin-console" in the field "Custom admin URL". When you have made this setting, you will notice that at the top of the plugin page, a message appears and tells you this:

Default (unsafe) | Lite mode

**WARNING:** The admin path is hidden from visitors. Use the custom login URL to login to admin

If you can't login, use this URL: **http:// /wp-login.php?hmw_disable=988603**
and all your changes are roll back to default

This message confirms that visitors to your site will no longer be able to access the back office of your WordPress site by going to the url *http://your_domaine/wp-admin*, they will get an error page (404).

The 2nd part of the message is very important, indeed, if for any reason, you can not access the back office of your site using its new url (in our example: http://your_domaine/admin-console), you can disable this setting by putting the indicated url (http://your_domaine/wp-login.php?hmw_disable=...) in your browser. Take the time to copy/paste this url before saving the settings of the "Hide My WP" plugin. I've never had a problem accessing the back office using this plugin, but if this happened to you, you'd be very happy to be able to restore access.

On the same principle, you will see that in its freemium version, the plugin also allows you to modify the url http://your_domaine/wp-login. Indeed, this url is the authentication url for all the users of your site. Its modification will have the same beneficial effects as the modification of the url http://your_domaine/wp-admin.

## Number 6: Hide the WordPress version you are using

The readme.html and license.txt files contain information about the version of WordPress used on your site and are easily accessible via the following urls:

*http://your_domaine/readme.html*
*http://your_domaine/license.txt*

The fact that a hacker knows the version of your WordPress instance can be problematic.

Indeed, as explained above, WordPress, like any software, has security holes. The complexity of software such as WordPress makes it impossible to guarantee that there are no security holes, even with experienced developers. That is why, as soon as a security hole is discovered, WordPress developers fix it and release an update (i.e., a new version) of WordPress. If the version of WordPress you are using is freely available on your site (which is of no use to the people who visit your site) and you are not using the latest version, then it means that known security vulnerabilities are still open on your site. By knowing the version of your WordPress instance, the hacker can immediately know which technique(s) to use to hack your site. Hackers can even use a software robot that will automate the search for WordPress instances of a given version and on which it will automatically exploit vulnerabilities not yet fixed in your version.

That is why you have to:

1) Update WordPress systematically
2) Hide your WordPress version from users

Security updates for a given version are done automatically in WordPress. However, major updates (e.g. from version 4.9 to version 5) are done manually. After a while, older versions of WordPress no longer benefit from security updates (as it becomes too costly for developers to publish patches for all versions of WordPress) and if you are using an older version, your site is vulnerable to attack via known and public vulnerabilities explained on the Internet. You must therefore scrupulously make sure to keep your WordPress instance up to date!

To hide the version of your WordPress instance, you must prevent access to the readme.html and license.txt files. To do this, the simplest way is to use the .htacess file which is at the root of your site.

Just add these lines at the end of the .htaccess file:

```
<files readme.html>
deny from all
</files>

<files license.txt>
deny from all
</files>
```

The version of your WordPress instance is also present in the HTML code of the pages on your site. To remove it, you have to modify the function.php file of your theme.

Just add this line at the end of the file function.php:

```
remove_action("wp_head", "wp_generator");
```

To be able to copy/paste these settings lines into your WordPress configuration files, send me an email to wpbonus@gmx.com, with "BONUS WP" in the subject line, you will receive an email containing all the settings and plugins recommended in this book.

## Number 7: Change database prefix

WordPress stores the settings, the users, and the content of your site (articles, pages, categories...) in a SQL database. The MySQL database is the most frequently used.

In an SQL database, data is organized in a table (like an excel table) containing columns and rows. In the world of databases, these tables are called "tables". Each table has a name and is used to store information.

SQL (Structured Query Language) is the language used to query the tables in a database. With an SQL query, it is possible to view/create/modify/delete data in tables. Here are some examples of SQL queries translated into english:

- In the user table, I want to get the email address of the user whose login is: admin (this is an example, in real life, you should not use the admin account)

- In the article table, change the last update date whose article ID is...

The WordPress engine (written in PHP) spends its time querying the database with SQL queries.

By default, WordPress creates multiple tables that all have the prefix "wp_".

Here is the list of WordPress tables:

- The wp_users and wp_usermeta tables: they contain the list of users with their main information
- The wp_posts and wp_postmeta tables: They contain the articles and pages of your WordPress site

- The wp_comments and wp_commentmeta tables: They contain the comments present on your site
- The wp_options table: It contains the settings for the WordPress installation and extensions (plugins)
- The tables wp_terms, wp_taxonomy, wp_term_relationships and wp_termmeta: They contain taxonomies, which allow to group articles according to typologies. Two taxonomies exist as standard: categories and labels
- The wp_link table: It contains all the links present on your site

By default, the names of all the tables in the WordPress site database are therefore known to everyone, including hackers! If the hackers manage to access the database, they can do whatever they want.

To do this, hackers use automated scripts that attempt SQL injections on known flaws in WordPress or certain WordPress plugins. Since the database contains virtually all the information on the site, these flaws can be exploited for one of the following reasons:

- Add content to the site (example: links to illegal sites)
- Add spam comments
- Destroy the content of the site
- Have the site planted
- Extract information (retrieve the list of emails of people who have left a comment)

SQL injection is a well known attack method. It consists in modifying an SQL query made by the WordPress engine to make it play another query than the one normally coded in the engine. The technique consists of injecting pieces of SQL code through a data entry form. By this technique, the hacker will succeed in querying your site's database without knowing the password (since he injects his own queries into those made by the WordPress engine). Not knowing the password of your database (because you put a very strong password on your database), the hacker cannot query the database to get the list of tables and their names, he has no choice but to attempt queries using the default names. By leaving the default table prefixes, you indirectly facilitate the work of these hackers.

To change the prefix of your database, there are two possible scenario:

1- You have not yet installed WordPress

It's the simplest scenario. During the installation, you just have to change the value "Table Prefix" which is by default "wp_", to the value you want, for example "ghtfk87_". For hackers, there is no way to know that the table prefix on your WordPress site is "ghtfk87_".

Below you should enter your database connection details. If you're not sure about these, contact your host.

| | | |
|---|---|---|
| **Database Name** | wordpress | The name of the database you want to use with WordPress. |
| **Username** | username | Your database username. |
| **Password** | password | Your database password. |
| **Database Host** | localhost | You should be able to get this info from your web host, if `localhost` doesn't work. |
| **Table Prefix** | wp_ | If you want to run multiple WordPress installations in a single database, change this. |

Submit

Don't choose a prefix starting with wp, because some hackers use scripts that look for all tables with names in the form wp* (i.e. wp followed by any character).

2- You already have an instance of WordPress installed

The operation is more complex but can be done very well with SQL queries executed via PHPMyAdmin. However, to make it easier, I recommend using the Brozzme DB PREFIX plugin. Good practice is, of course, to back up your database before making any structuring changes to it. The Brozzme Db Prefix plugin is safe and used by many sites, however, a bug can still occur. That's why, it is necessary to make a backup of the database and of the wp-config.php file beforehand.

For the file wp-confg.php, you just have to connect to your server by FTP (I recommend the open source tool Filezilla) and download the file wp-config.php on your computer.

To back up your database, we will use the BackWPup plugin, we will see later how to use this plugin in more detail. Here we will just use it to download a backup of the database, so that we can restore it in case of a problem. Backing up your site (and knowing how to restore it) is essential (for example to be able to put your site back in order after an attack) and this point will be detailed later in the book.

As soon as the BackWPup plugin is installed and activated, in the WordPress back office, on the left side panel, you will see a new section: "BackWPup". In this section, click on "Dashboard", then you will see a box entitled "Backup in one click", click on the button located inside and entitled "Download a backup of the database", you will then download a .sql file, keep it preciously, it contains a complete export of the database of your site (all your articles, all pages, users, comments ...).

With the database and the wp-config.php file saved, we will be able to use the Brozzme DB PREFIX plugin to change the prefix of the tables in your database. This plugin is very easy to use. From the new Brozzme section on the left side panel, click on "DB PREFIX", you will see the following screen:

Database Prefix Settings

Before execute this plugin:

Make sure your `wp-config.php` file is writable.
And check the database has ALTER rights.

Existing Prefix: * wpetd_ ex: wp_
New Prefix: * kjw_ ex: uniquekey_
Allowed characters: all latin alphanumeric as well as the _ (underscore).

Change DB Prefix

Before clicking on the "Change Database Prefix" button, you have to check that, the wp-config.php file is indeed modifiable. i.e. that WordPress can modify it, because otherwise the prefix change will be incomplete (only in the database but not in the wp-config.php file) and you will have to restore the database (if necessary, I explain how to do it, in appendix, at the end of the book).

To check this point, the easiest way is to connect by FTP on your hosting (I repeat myself, but I advise you to use FileZilla):

- To right-click on the wp-config.php file,
- To left-click on the "File Access Rights" item in the context menu,
- Check that the "Read" and "Write" boxes are checked in the Owner and Group permissions. If not, you have to tick them off.

Once this check is done, you just have to enter your new prefix (don't choose a prefix starting with wp, because some hackers use scripts that look for all tables whose name is of the form wp*) and click on the "Change DB Prefix" button.

## Number 8: Hide the true IP address of your site with CloudFlare

Your WordPress instance is installed on a server, which, like all servers connected to the Internet, has a unique IP address.

**By knowing the IP address of the server hosting your website, hackers can try to exploit vulnerabilities in the server** (for example, if the server's operating system is not up to date or does not comply with good IT security practices). Hiding the IP address of your website is therefore an interesting protection.

Before explaining how to do this, a little reminder is in order:

When you go to a website, you type the url of the site in your favourite browser and the site is displayed. This operation, which takes place almost instantaneously for you, can actually be broken down as follows:

1) Your browser queries the Internet DNS by submitting the domain of the website you entered and the DNS returns the IP address of the server hosting your website.

2) Your browser queries the IP from the DNS and retrieves an HTML page that it displays on your screen.

The DNS (Domain Name System) is a service for mapping a domain name to an IP address. The sequencing described is exactly the same as when you call "Mom" on your phone. Your phone uses your contact list to call the phone number corresponding to the "Mom" contact. You don't have to know Mom's phone number by heart.

Here's a table with the analogy:

| Web navigator | Telephone |
| --- | --- |
| url of the website you want to assess | The name of the person you wish to call |
| The browser queries the DNS for the IP address of the requested website. | The phone queries the contact book for the phone number of the contact you want to call. |
| The browser queries the retrieved IP and displays the returned HTML page. | The phone calls the phone number and allows you to speak with your contact. |

To go to your website, your browser must therefore know the IP address of your website. It therefore seems contradictory to want to mask the IP of its website while allowing Internet users to access it, it is like wanting to receive mail without wanting to give your postal address.

Yet it is possible! Simply subscribe to a P.O. Box at the Post Office. The Post Office receives your mail in the P.O. Box and forwards it to your postal address. In this mode of operation, only the Post Office knows your home address. As the people who write to you do not know your home address, they will never be able to come to your home to try to pick your lock!

In computing, for websites, the equivalent of a post office box is called a CDN (content delivery network) or réseau de diffusion de contenu (RDC) in French. The principle of the CDN is that it has a copy of your website (technically, your website will be cached in the CDN's cache memory) and that it is this copy that it sends to the Internet user's browser. In this way, the Internet user's browser displays your website without knowing its IP address, just like a P.O. box that hides your postal address. By delegating the DNS of your website to a CDN, it will be the IP address of the CDN that will be sent to the browser of Internet users and not the IP address of your website. Your website will then be requested only by the SSC. Indeed, the latter will still request your site to retrieve the dynamic contents of your site (articles, pages...) and will use its copy for static contents (which change very little, such as images and CSS files).

A CDN consists of a set of servers located in different geographical locations and connected to each other via the Internet. It sends your web pages from the server closest to each visitor. Thus, if your american site is requested by a french Internet user, it will be the CDN server in France that will send him the copy of your website. The display of your website will be much faster (because there is no need for the request to cross the Atlantic to be displayed in the browser) and the IP address of your website will be masked.

The icing on the cake, by using a CDN, your site will be able to handle an increase in traffic much better. Thus, in the event of a sudden surge in the number of pages viewed on your site or blog (in case one of your articles goes viral), your site will remain online and will not fall under the weight of numerous solicitations (which would happen without going through a CDN).

The CDN CloudFlare, in its free offer (freemium) will thus allow both to secure your site (by masking its IP) and to make it faster by significantly improving its display time.

Once your site is integrated into CloudFlare, its traffic will be routed through their network. As a CDN, CloudFlare will automatically optimize your website, both on load time and performance. Most importantly, thanks to its regularly updated database of major attack types, CloudFlare will recognize and block many threats and attacks before they even reach your site.

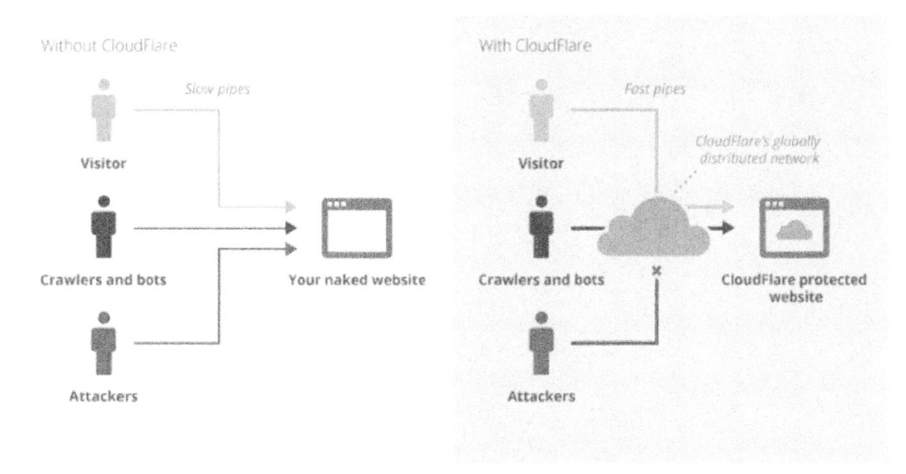


As usual, if you want more features, you can always switch to the pro offer, but the free service is enough for most of us.

At the time of writing this book, CloudFlare is used by 10% of the world's websites and every day 20,000 new websites subscribe to it.

Here are all the steps to set up CloudFlare:

1) Create an account on cloudflare.com
I will not go into more detail about this step because it is not difficult. It is simply a matter of creating an account for yourself by filling out the forms displayed.

2) Once logged in with your newly created account, click on "Add Site" in the top right corner.

3) Enter your domain in the popup that appears and then validate.

4) CloudFlare will then warn you that it will query the DNS to retrieve useful information, so you won't have to re-enter it.

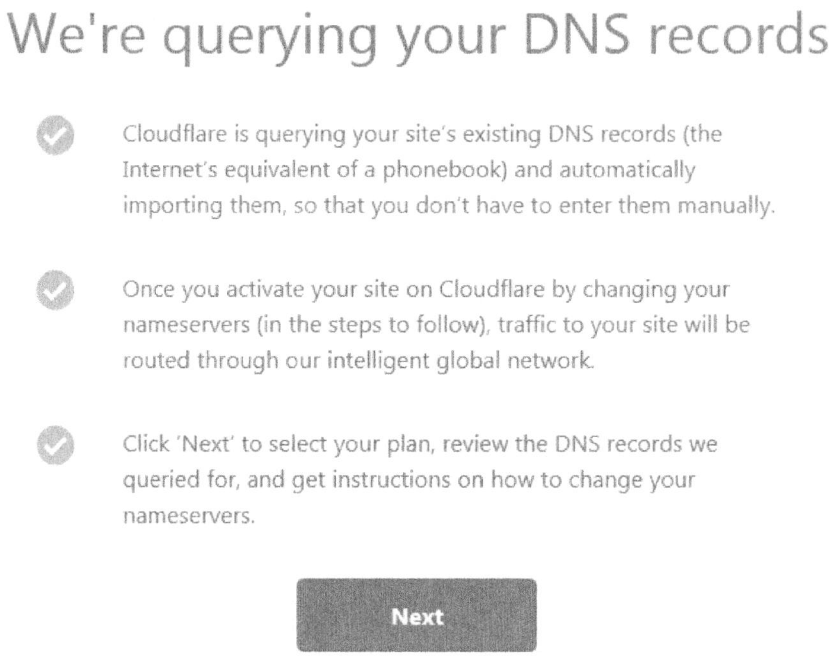

# We're querying your DNS records

- Cloudflare is querying your site's existing DNS records (the Internet's equivalent of a phonebook) and automatically importing them, so that you don't have to enter them manually.
- Once you activate your site on Cloudflare by changing your nameservers (in the steps to follow), traffic to your site will be routed through our intelligent global network.
- Click 'Next' to select your plan, review the DNS records we queried for, and get instructions on how to change your nameservers.

Next

5) As requested, click "Next"

6) Choose the "FREE" plan at $0 per month then click on "Confirm Plan" and confirm your choice

7) CloudFlare will show you the DNS records found, just click "Continue" at the bottom of the screen

8) CloudFlare will then tell you which CloudFlare DNS to use instead of the DNS you have been using so far. You may have the following screen:

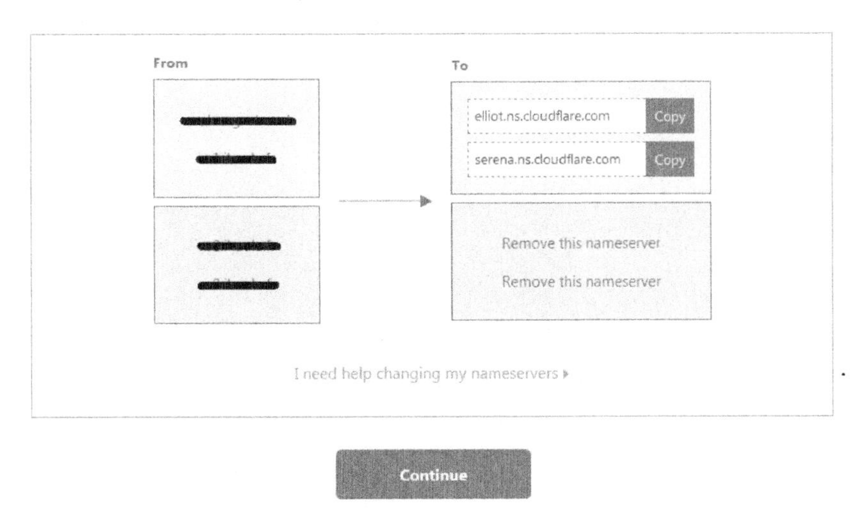


The principle is to set up the DNS of the host of your WordPress site so that it sends all the requests made to it to an external DNS, in this case that of CloudFlare.

CloudFlare has 2 DNS servers (at the time of writing this book: Elliot and Serena):

- If your hosting provider also has 2 DNS servers, they will have to be replaced by the 2 DNS of CloudFlare,

- If your host has more than 2 DNS servers, you will have to replace the first 2 and remove the others (in the screenshot above, the original site relies on 4 DNS servers, the first 2 should be replaced by elliot and serena and the last 2 should be removed).

To change your web host's DNS servers, you will need to log in to your web host's site and change the DNS server settings as specified by CloudFlare (here, replace your web host's DNS servers with elliot.ns.cloudflare.com and serena.ns.cloudflare.com servers). Until you make these changes, CloudFlare will not be used to access your WordPress site.

To change the DNS servers of your hosting provider, CloudFlare even offers tutorials to follow, to access them, just click on "I need help changing my nameservers".

How do you know if CloudFlare is active for your domain name?
Once you have replaced the names of the DNS servers of your hosting provider with those of CloudFlare, the change will be effective between 10 minutes and 48 hours depending on the propagation of the changes on all the Internet DNS servers in the world.

To know if the change is effective and your site is now protected by CloudFlare, simply go to your CloudFlare account home page.

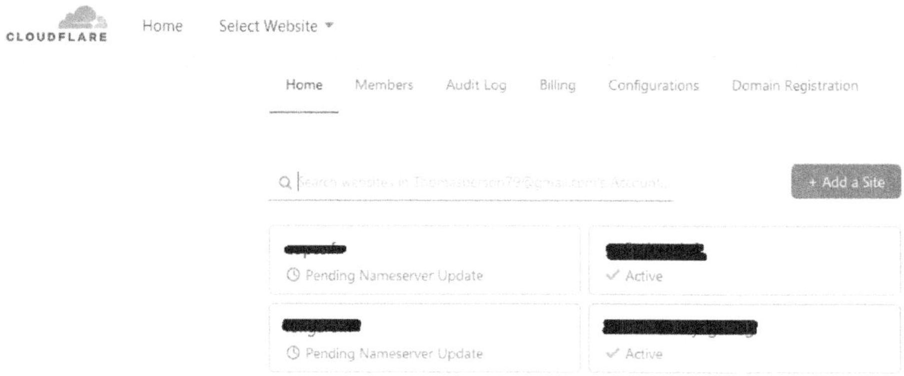

CLOUDFLARE
Home
Select Website
Home
Members
Audit Log
Billing
Configurations
Domain Registration
+ Add a Site
Pending Nameserver Update
Active
Pending Nameserver Update
Active

Your domain is managed by CloudFlare when it is subtitled "Active", if the mention "Pending Nameserver Update" appears, it means that your DNS has not yet switched to CloudFlare, then you have to wait. Switching your site to Cloudflare will be done hot (i.e. without interrupting the operation of your website) and therefore without impact for users who will continue to access your website without knowing that they are using CloudFlare.

Installing the CloudFlare plugin on WordPress:

Since CloudFlare is now enabled for your website, visitors to your website don't know it, but they request CloudFlare servers and CloudFlare responds to them. This operation has many advantages (mentioned above), but only one drawback: you can no longer know the IPs of your visitors. This is very annoying because you probably use Google Analytics to track your site traffic, and via CloudFlare, all your visitors will have the IP of the CloudFlare server closest to them. Your stats will be completely wrong!

To remedy this, there is a solution: Install the CloudFlare plugin for WordPress.

This plugin retrieves the IP of your visitors, which CloudFlare adds in the http header (I won't go into more technical details, because it's not necessary), so that you can know the IP of your visitors and thus continue to have real and exploitable statistics of frequentation. Moreover, if you don't install this plugin, it will be the IP of CloudFlare that will be associated with all visitors to your site and any comments they might leave. The CloudFlare plugin also has other handy features including the ability to automatically update the CloudFlare cache (the cache is the technical term to describe the copy of your website that CloudFlare has in memory and that it sends to your visitors) when you publish new content. Anyway, you got it: You have to install the CloudFlare plugin!

To install the CloudFlare plugin, simply type "CloudFlare" in the "Add Plugin" section of WordPress. When the plugin is installed, you will need to authenticate your CloudFlare account on the plugin. To do this, when you go to the settings, you will have the choice of creating an account or using your account. As we have just created a CloudFlare account and set it up for your site, you will need to use the "Have an account already? Sign in here. ».

Then, the plugin will ask you the email you used to create your CloudFlare account and your API Key. The Key API can be obtained from cloudflare.com. To do this, simply log into your CloudFlare account, then in the top right-hand corner click on "My profile", then go to the "Global API Key" section, click on "View" and click on "View".

API Keys
Keys used to access Cloudflare APIs.

Global API Key Change View
Origin CA Key Change View
Help ▸

copy and paste it into the settings of your CloudFlare plugin.

In the configuration of your CloudFlare plugin, you'll have to set the "Automatic Cache Management" parameter to "On".

As explained earlier, this setting is important because as soon as you add/modify/delete content, the CloudFlare "cache" (i.e. its internal copy of your website that it uses to respond quickly to users) will be deleted and recreated, so that there is no difference between the copy served by CloudFlare to users and your website.

## Number 9: Get your site to HTTPS with CloudFlare

HTTPS (for Hypertext Transfer Protocol Secure) is the secure version of the HTTP protocol. By using the HTTPS protocol, the data exchanged between the browser and the website becomes encrypted, making it unreadable any data intercepted by a hacker.

To use a comparison with mail sent by post: a neighbour can easily intercept your mail (by forcing your mailbox or pretending to be you with a replacement postman), read your mail, put it back in the envelope, and put the envelope in your mailbox. You'll never know that your neighbor read your mail. If your mail was sent using HTTPS, the contents of the envelope would be incomprehensible to your neighbor, because it would be encrypted knowing that only you have the decryption key.

This is why the HTTPS protocol is mainly used by e-commerce and banking sites. The objective is to guarantee the customers of these sites the protection of their personal data, such as their access identifiers or their credit card number. All sites that support financial transactions have long since adopted this security protocol.

HTTPS helps make your site safer for visitors to your site. That is his main interest. By using HTTPS, you limit the risks that a malicious person will intercept the data transmitted by the Internet user on your website and more generally all information transmitted between the browser and the server of your site.

You could say to yourself that since your site is a blog for the public, without an online payment system, the switch to HTTPS does not concern you. It is not. Indeed, the HTTP protocol has certainly enabled the development of the Web, but it is very vulnerable. Indeed, it allows anyone who controls the network you use (Wifi from hotels, Internet cafes, co-working spaces and of course your Internet service provider) to modify the content of the http sites you consult, without you being aware of it.

As Troy Hunt, security analyst at Microsoft, reminds us, here are some examples of possible threats to HTTP sites:

- Inserting advertisements (or other content) that are not on the original website

- Injecting invisible software that undermines crypto-money with your pc for the benefit of a third party (for the record the mining of crypto-money makes money), as practiced by a Starbucks store in Argentina in 2017 via the wifi access it provided to their customers

- Redirect visitors to fake websites (being a carbon copy of the original site) with a technique called DNS hijacking. The user who thinks he is on the site he asked for (since it is the url present in his browser), enters his login/password and the hackers retrieve it (since in reality, the user has entered this information on their site) and can use it on the real site. (Good practice: Never authenticate yourself on an http website, as it is not secure)

All these malicious acts are impossible for HTTPS sites.

This observation has moreover pushed Google to alert Internet users using its browser (Google Chrome) when they go to unsecured sites. Therefore, that today, if your website contains a form and is not in HTTPS, Chrome will present a security alert to Internet users. Their goal is to make the whole web secure in HTTPS, this behaviour of the Chrome browser is a nice incentive!

Another incentive, and not the least, **Google has decided to give HTTPS websites an advantage in its search engine results ranking**. So to optimize your natural referencing, you should also put your site in HTTPS! In this context, at the time of writing this book, more than 50% of the web is now in HTTPS, if your website is not yet HTTPS, it is urgent to go there!

CloudFlare allows, very simply, to pass any website (not only sites using WordPress) in HTTPS. There are other ways, such as with "Let's encrypt" or via your web host (if they offer SSL certificates) but using CloudFlare is easier, and since we just put your website behind CloudFlare, you might as well take advantage of it!

Here's how to do it:

1) On your CloudFlare dashboard, in the "Crypto" tab, under "SSL", select "Flexible"

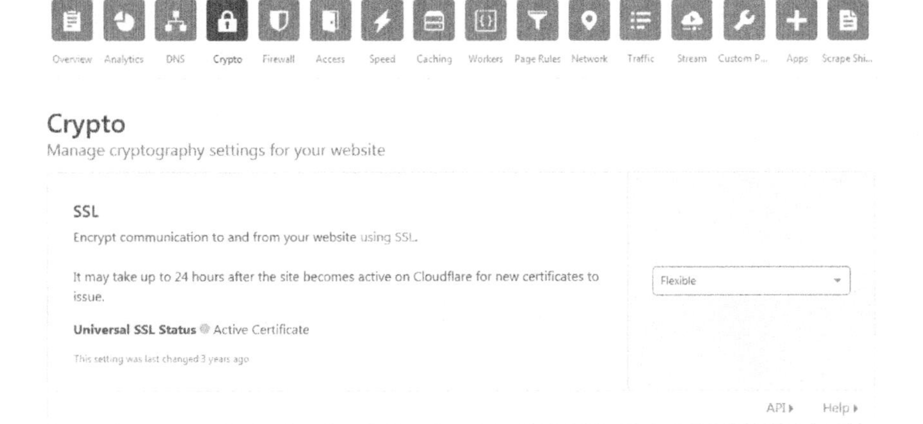


2) Go check on a browser that your site is now accessible in http and https. To do this, nothing could be simpler: add an "s" to the url of your site https://<url of your site>
As long as your site is not accessible in https, do not proceed to step 3

3) This step aims to automatically refer users to the https version of your site when they request your site in http.

This point is doubly important:

- If your site is always accessible in http, it is not secured.
- Your website is accessible via two urls (http and https) and Google does not like this (it considers, wrongly, that your website has been duplicated) and will degrade its natural SEO

In the "Crypto" tab, set the following settings to "On":

- Always Use HTTPS
- Automatic HTTPS Rewrites

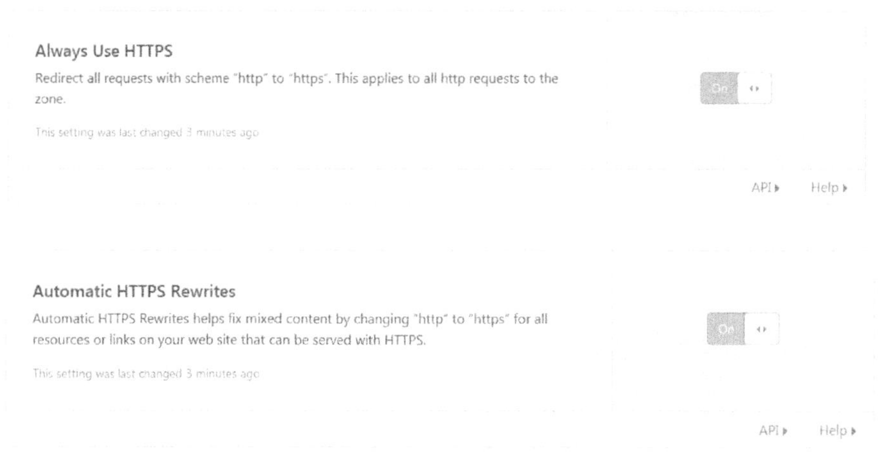


The change will take a few seconds to become active.

In this way, whatever the url requested by an Internet user, he will be automatically redirected to the HTTPS version of the site.

After switching to HTTPS, your browser may indicate that your site is not completely secure, with an exclamation mark added to the HTTPS symbol.

Site Information for

 Connection not secure

Parts of this page are not secure (such as images).

Permissions

You have not granted this site any special permissions.

Clear Cookies and Site Data...

On the pages or articles concerned, this means that you have used the full url of an image instead of using the relative url.

For example, for an image "logo.jpg", the complete (or absolute) url of your image would be of the type:
*http://<your_domain>/wp-content/upload/logo.jpg*

So, your site now in HTTPS would contain an HTTP image, which would cause an alert message on your browser.
It will then have to be replaced by the relative url (i.e. one that does not include the domain name or the protocol), which will be in the form:
/wp-content/upload/logo.jpg

The image will still be displayed but there will no longer be an alert in the browser, in fact, as the protocol is not specified, the browser will use the same protocol as the one entered in the browser bar, it will request the image via *https://<your_domain>/wp-content/upload/logo.jpg*.

The HTTPS page will no longer contain HTTP images and therefore the alert will disappear.

The use of relative url for calls to local resources (eg images, css, javascript ...) (ie on your own site) is a good practice, I advise you to follow this good practice, because Google sanctions the natural referencing of HTTPS sites containing elements in HTTP and for your visitors, a security alert in the browser, it is not very serious.

## Number 10: Disabling the XML-RPC interface

XML-RPC is a standard protocol that allows two sites to talk to each other. It is widely used on the Internet and not only by WordPress.

WordPress implements this standard protocol (via the "xmlrpc.php" file that is at the root of your site) because it allows the development of third party services, plugins or task automation via IFTTT or Zapier (which use XML-RPC to interact with WordPress).

This protocol is also used by some plugins such as WordPress Mobile App, JetPack, BuddyPress... or when you create an article from an external application (e.g. Windows Live Writer).

This protocol is very practical but diverted from its normal use, it can greatly facilitate the life of hackers. Indeed, in a single XML-RPC call, a hacker can send a high number of authentication attempts, for example to try to find your admin password (as we saw previously). This technique is a hundred times more effective than "normal" brute force attack attempts.

Hackers can also make your site unreachable (denial of service attack) by sending a large number of requests to the server and thus cripple it. This form of attack is called HTTP Flood Attack, because your site is flooded with HTTP requests.

If you do not really need to connect to WordPress via external applications, you may want to disable XML-RPC, to avoid attacks from external applications.

To do this, we will use the iThemes Security plugin that we have already installed to protect ourselves from brute force attacks.

Just go to "WordPress tweaks", click on "Configure settings" and then you will see the XML-RPC option:

XML-RPC

WordPress' XML-RPC feature allows external services to access and modify content on the site. Common example of services that make use of XML-RPC are the Jetpack plugin, the WordPress mobile app, and pingbacks. If the site does not use a service that requires XML-RPC, select the "Disable XML-RPC" setting as disabling XML-RPC prevents attackers from using the feature to attack the site.

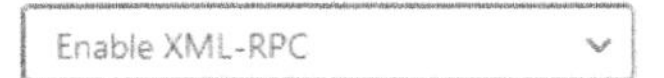


- **Disable XML-RPC** - XML-RPC is disabled on the site. This setting is highly recommended if Jetpack, the WordPress mobile app, pingbacks, and other services that use XML-RPC are not used.
- **Disable Pingbacks** - Only disable pingbacks. Other XML-RPC features will work as normal. Select this setting if you require features such as Jetpack or the WordPress Mobile app.
- **Enable XML-RPC** - XML-RPC is fully enabled and will function as normal. Use this setting only if the site must have unrestricted use of XML-RPC.

Multiple Authentication Attempts per XML-RPC Request

WordPress' XML-RPC feature allows hundreds of username and password guesses per request. Use the recommended "Block" setting below to prevent attackers from exploiting this feature.

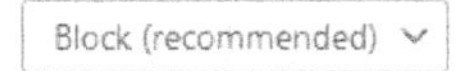


- **Block** - Blocks XML-RPC requests that contain multiple login attempts. This setting is highly recommended.
- **Allow** - Allows XML-RPC requests that contain multiple login attempts. Only use this setting if a service requires it.

As indicated, it is recommended to disable XML-RPC. I advise you to do so, and then check that your site is still working well. If necessary, if you notice a plugin malfunction, you can change this setting to "Disable pings".

## Number 11: Blocking navigation of your WordPress folders

Browsing your website folders can be used by hackers to find out if your site contains files with known vulnerabilities or by others to view your files, copy your images, learn about your website structure and other information. For this reason, it is strongly recommended that you disable indexing and browsing in the directories of your WordPress site.

Some security plugins allow you to easily disable this navigation in directories, but you can also add this little piece of code in the .htaccess file located at the root of your site:

```
Options -Indexes
```

This will prevent the curious from going further in their navigation and will simply return a 403 error page: Access to the file requires authorization.

For this protection, which is simple to set up without a plugin, I recommend that you do not use a plugin, as it is good practice to limit the number of active plugins on a WordPress instance. Indeed too many plugins can reduce the performance of your site and even cause strange behaviors (because some plugins are incompatible and cannot work together).

## Number 12: Disabling the File Editor

This is a basic security principle that should be applied to all your WordPress sites. As a basic WordPress package, WordPress comes with a built-in code editor that allows you to edit your theme and plugin files from your administration area on the Appearance > Editor tab. If a hacker manages to connect to the back office of your WordPress site, via the editor, he will be able to modify the core of your website. This security risk can be completely blocked by disabling the editor.

So here's how to disable the plugin and theme editor from WordPress features, copy the line below into the wp-config.php file at the root of your FTP:

```
define('DISALLOW_FILE_EDIT', true );
```

When creating a site, it can be convenient to be able to edit files using the WordPress editor. In this case, it is quite possible to wait until your site is finished before disabling the file editor. If one day, you need to make changes to some files, you can do so, either via FTP, with Filezilla, or by temporarily reactivating the WordPress editor (by deleting or better by commenting the concerned line in the wp-config.php file).

## Number 13: Disabling the execution of PHP files in certain WordPress directories

WordPress generates HTML pages that are presented in your web browser via scripts using the PHP language. The PHP language is widely used to make dynamic websites (where the content displayed is based on a database as opposed to a static website). When a PHP page is called (for example, when you go to a PHP page with your browser), the actions encoded in the scripts are executed on the server where the website is hosted and the result of this execution is the HTML code that is presented to you in your browser (which you can see by right-clicking, displaying the source code).

The PHP language can for example be used to query a database and display the result of the query in the browser. It can also be used to encode malicious actions (recovering all user passwords from the database, deleting all folders from the website, changing text on the pages of the site...).

If a hacker manages to upload a malicious PHP page to your site (for example because on your site, you can send an attachment), he will just have to call him (by putting his url in his browser) to do a lot of damage to your website.

The WordPress engine relies on PHP files in specific directories and of course does not use PHP files in the directory where uploads are stored (/wp-content/uploads/). In other words, there aren't supposed to be any PHP pages in your uploads folder, so if there are, they are probably malicious. In order to stem the threat at its root, it is entirely possible to prohibit PHP scripts from running when they are in the /wp-content/uploads directory.

You will then have to create a file named .htaccess that you will place at the root of your /wp-content/uploads/ directory via FTP and that will contain these lines of code:

```
<Files *.php>
deny from all
</Files>
```

In order not to overload your WordPress site with plugin and thus slow it down, I prefer to use only the iTheme plugin for security and proceed by changing the site configuration files for the other protection actions. However, be aware that this feature is also offered by some security plugins such as Sucuri, Wordfenc or SecuPress.

## Number 14: The Role of Hosting Your WordPress website

Your host plays one of the most important roles in securing your WordPress site. A good hosting provider needs to take extra steps to protect your site from common threats.

In the case of shared hosting, you share a server with other sites that can potentially pose a threat to yours: this increases the risk of contamination or usurpation if your "neighbour" is the malicious type!

Be sure to choose quality hosting providers that offer additional features such as automatic backups, as well as advanced security configurations (anti DDoS, firewall ...).

Hosting companies specialized in WordPress are also a good option to consider (non-exhaustive list):

- ThemeCloud
- WP Server
- o2switch
- WP Engine
- Flywheel
- WPX
- Kinsta
- Bluehost

## Number 15: Moving your PhpMyAdmin

This web application allows you to manage your databases. If it is accessible at the following address: /monsite.com/phpmyadmin, then it is strongly recommended to move it because hackers will try to access it via this url.

To do so, you will need to contact your hosting provider to explain the procedure to follow. However, if you use a special WordPress hosting (shared or dedicated), the hosters do their job properly and do not put PhpMyAdmin in direct access. Most often, you have to go through the host's dashboard (by being previously connected with your account) to access it.

## Number 16: Be careful with the themes and plugins you install

A good practice is to install as few plugins as possible and to install **only regularly updated plugins offered by WordPress** via its plugin search interface. Indeed, many attacks on WordPress are based on plugins containing security holes. As plugins are developed by third parties (not by the developers of the WordPress engine), IT security issues are not necessarily taken into account (unlike WordPress which is very regularly updated for these reasons).

Similarly, some themes may contain security holes. To avoid this, give priority to regularly updated themes and keep them up-to-date!

Finally, you may have already come across websites offering free downloads of premium themes and plugins (normally paid for). It is absolutely not necessary to download these files because the hackers who propose these files, could have voluntarily added security flaws to the code of the themes or plugins that they propose to you. This way, they will be able to hack your site very easily.

## Number 17: Make backups

Backups of the system should be made at least once a week so that you can restore your site in case of hacking. Better safe than sorry!

As explained above, your WordPress instance stores its data (content and settings) in its database and in php configuration files. Here again, we will use the BackWPup plugin. It offers backup planning as standard.

If your hosting provider offers backups (you should not choose a hosting provider that does not offer backups), enable automatic backup and it will not be necessary to schedule backups on BackWPup.

If your web host does not offer any or if you are not satisfied with the frequency of the web host's backups, you will need to use the BackWPup plugin.

Here's how to do it:

In the plugin menu, click on "Add a task"
In the "General" tab:

- Name the task: "Complete site"
- Check the following 3 boxes:
  - Backing up the database
  - File Backup
  - List of installed extensions
- Leave the archive name as it is by default.
- Check the "Zip" box for the archive format

For the destination of the task, the easiest way is to choose "in the local folder" where you can access the backup zipped files directly on the disk space of your hosting by FTP with Filezilla.
In the "Database" tab:

- Check all tables
- File compression: None

In the "Files" tab:

- Leave the default setting

In the "Extensions" tab:

- Leave the default setting

A txt file containing the list of your plugins will be present in each backup. If you restore on an empty WordPress instance, you will know exactly which plugins to reinstall.
In the tab " to: Folder ":

- You will be able to specify the folder to which you want to save your backups, it is in this folder that you will go with FileZilla to retrieve your backups.

- The number of backups to be kept should also be indicated. If you put 10, you will be able to access the last ten backups. The goal is not to saturate the disk space of your hosting.

In the "Planning" tab:

- Check "With WordPress cron"
- Leave planner type set to "basic"
- Choose the frequency of your backup

The frequency of the backups is to be chosen according to the frequency of update of the content of your site. If your site changes twice a year, an automatic daily backup will not be necessary.

To check that your backup is correctly set up, you can go to the "Tasks" section of the plugin menu, you will see your "Complete site" task and its schedule.

In the appendix, you will find how to restore your WordPress site.

# Chapter 2: How do you recognize a hacked WordPress site?

## How to monitor your site?

To track your website activity (and detect abnormal situations, proof of probable hacking), you must absolutely link your WordPress site to Google Search Console, Google Analytics and the UpTime Robot tool.

All the websites (whether they are under WordPress or not) for which I am responsible use these two (free) services from Google and UpTime Robot (in its free version). These indispensable tools will allow you to obtain your site's health record by revealing the metrics and trends of navigation on your site: the number of visitors to your site, the most visited urls, the time that Internet users spend on your site, the response time of your site... Uptime Robot will alert you by email if your website no longer responds, so you will be informed very quickly (before your users send it back to you) if your site is unavailable.

Here's how to set up these three tools:

Uptime Robot

All you have to do is create an account on https://uptimerobot.com. When you are authenticated, simply click the "Add New Monitor" button and configure the settings as follows:

- Monitor type: HTTP(s)
- Friendly Name: This will be the name of your site as it will appear on the dashboard.

- URL (or IP): Put the url of your site for which you want Uptime Robot to check the display time every five minutes. For my sites, I put the home page.
- Monitoring Interval: by default it is set to Every 5 minutes, I leave this value.

Google Analytics

This service from Google allows you to finely track:

- The number of visitors to your website,
- Where do visitors to your site come from (search engine, social networks...), in what geographical area your visitors are,
- To find out which pages are the most visited,
- To follow the response times of your site (important point if you want your visitors to stay on your site, it must be fast).

To take advantage of this great free service, go to https://analytics.google.com

At the bottom right, click on the cogwheel labeled "Administration" and then create a property (a property represents your website). You will then have to fill in the requested information such as your website url, the category of your website, the time zone that will be used for the reports and click on the "Get Tracking ID" button. You will have to wait a few seconds for the site to reload and your Tracking ID will be displayed, it will be in the form: UA-xxxxxxxx.

You'll just have to copy/paste it because we'll put it on your WordPress site so that everything that happens on your site is sent to Google Analytics. Don't worry, this will not penalize the performance of your website but will give you access to very interesting reports (on Google Analytics) about the traffic of your site.

On your WordPress site, if you are using a professional theme, in the theme setup there is likely to be a section where you can copy and paste your Google Analytics tracking ID. This section is often referred to as SEO or Analytics. If you don't have this option, just install the "Google Analytics Dashboard for WP" plugin. You will then have to allow the plugin to connect to your Google Account, via the generation of a single-use code, which you will get via a link in the plugin settings. The plugin will then automatically retrieve your tracking ID and activate it on your site.

Google Search Console

Google Search Console allows the administrators of a website to check how often Google's crawlers index their website. It allows site administrators to track and manage the site's Google referencing, including :

- Know on a daily basis the keywords that led to a visit to their site
- If necessary, prevent certain urls from appearing in Google search results
- Improving and optimizing the content of certain HTML tags that are widely used for natural referencing (or SEO for Search Engine Optimization)

It is a powerful and essential tool for those who want to optimize their search engine optimization on the web. However, since this book is not about SEO (surely my next book), I will not dwell on these aspects and focus on the uses of Google Search Console in case of an attack.

To declare your site on Google Search Console:

- Log on to https://search.google.com/search-console
- Click on Start
- A popup opens to welcome you, click on the "Start" button.
- Then at the top left, click on "Add Property", then choose "Prefix url" and enter the url of your website.
- You will then have to prove to Google that you are indeed the site administrator.

Google Search Console offers several methods of verification for this purpose. I recommend the "HTML File" method, which I think is the simplest. If Google doesn't offer this method by default, you can choose it from a list at the bottom of the form. This method consists in downloading a small HTML file generated by Google and putting this file at the root of your website.

The file generated by Google will have a name like googlexxxxxxxxxxx.html, it is just a matter of putting this file at the root of your site so that Google can request it via http(s)://<your domain>/googlexxxxxxxxxxx.html. This will prove to Google that you are indeed the site administrator, because only an administrator can add files to the root of the site. So you will have to download the file generated by Google and upload it to your site via FTP, nothing very complicated. Your site will then be successfully declared on Google Search Console.

## How do you know if your site has been hacked?

During a hack, webmasters are often the last to know!

Here are some indicators that can help you avoid this:

The traffic of your site becomes huge and your site becomes very slow (the response time of your site exceeds 5 seconds):
Your site may be a victim of a denial of service attack, your site is bombarded with requests by hackers who want to saturate the server hosting your site so that it becomes unavailable. Having CloudFlare on the front end of your site protects you from many of these types of attacks and normally your hosting provider must also be equipped to detect this type of attack and block the attacking IPs (hence the importance of choosing your hosting provider carefully). You can also see this if one of your publications goes viral and the requests made on your site become too numerous. You will then have to upgrade your hosting to take into account your new popularity.

The traffic of your site drops sharply:
There can be several reasons for this such as your server being inaccessible, a bug during a WordPress update, a penalty from Google on SEO (Search Engine Optimization) that would remove your site from the first page of Google results, but also a hacking, which would have simply erased your website (hence the importance of making backups).

Your site changes its appearance:
It's pretty obvious at first glance. If your site suddenly changes its appearance, you should log in to your administration to take a closer look at it. You can also go to the Search Console to see if certain information can be given to you.

A message appears in Google results when searching for your site ("it is possible that this site has been hacked"):
If such a message is associated with your search result, you should quickly go to the Google Search Console and follow Google's recommendations.

The clickable links on your site lead the Internet user to sites that are unknown to you:
It's one of the most common hackings. A hacker uses your site to send your visitors to spammy sites (viagra, pornography etc.). If this happens to you, there is no doubt your site has been hacked! If your site contains few pages or articles, you can go to each of them and restore a previous version (native WordPress function). If your site contains many pages and articles, it is best to restore a complete backup of the site.

You discover a suspicious user:
You don't always think about it, but it is important to go to the Users tab> All Users tab from time to time to watch for new users popping up out of nowhere. Indeed, it is a rather well-known hack: a hacker sneaks in discreetly and creates a new Administrator. If this is your case, try to delete it and find out where the flaw comes from, if it was entered once, it may come back (for example, change your Administrator password, change the default url for back office access if you haven't already done so...).

You can't log in anymore:
This can come from a variety of sources. It could be that one of your security plugins has a bug, that you have simply forgotten your password and/or login, or it could be that a malicious person has managed to infiltrate your backoffice and delete your user account. If this is the case, from the host's site, you will need to restore a backup of your website, and then strengthen the protection of your site by applying the advice in this book.

There are still several cases that can tell you that your WordPress site has been hacked, but most often, it is the users (visitors to your site) who can inform you, because their browser has warned them (this is often the case with Chrome). In this case, if you are regularly notified of such alerts, don't take it lightly and check the points mentioned above.

## What to do after a WordPress hack?

The vast majority of attacks rely on the flaws of poorly secured plugins. Indeed, as explained above, some plugins are never updated for security patches. If your WordPress site has been hacked, you can start by disabling your plugins one at a time, so that you can find the plugin that the hacker used to enter your site. If, when a plugin is deactivated, your site regains a normal aspect, it is this plugin that will have allowed the hacking. It will then have to be replaced by another more secure plugin.

If the methodical deactivation of the plugins did not restore your site, then you will have to restore a previous version of your website. In order to do this, you must have set up automatic and regular backups beforehand. This solution is radical (provided you have a healthy backup of your site) but has the disadvantage of a possible loss of content. Indeed, if you find that your site has been hacked for two weeks, you will have to restore an earlier version and therefore lose all publications that have been made since. If you can still access your pages and articles, export them in XML format, you will be able to re-import them after the restoration so you don't lose any content. If you can no longer access your pages and articles, you will have to restore the last healthy version of your site.

It is not always easy to find the date from which your site was hacked, so you will have to try each of your backups starting with the most recent (by restoring them and checking the appearance of your restored site), until you find the healthy backup (i.e. a backup made of your site before it was hacked). By doing so, you will find the most recent healthy backup, i.e. the one with which you will lose the least amount of content. Many WordPress users don't realize the importance of backups and security for their site until it is hacked. Better safe than sorry!

You can use the restoration features offered by your hosting provider or take care of the restoration yourself. In this second hypothesis, it will be necessary to restore both the database and all the files on the server (I explain how to restore backups made with BackWPup in the appendix).

In some cases, you may also want to consider re-registering your site in the Search Console so that Google will reconsider it as a healthy site.

## In conclusion....

The Internet is a vast playground - or rather hunting ground - for hackers. Alongside the major computer attacks that we have all now heard about, less organised and less visible cybercrime is very active, and it is probably this that we as individuals should be most wary of. Surfing the Internet, using applications, social networks, online payment methods, all this should no longer be done without being aware of the risks involved. It is indeed now important to instil a true culture of caution related to digital uses, for young and old alike.

As a WordPress webmaster, you may be the target of large-scale attacks by software robots that will automatically test the major types of attacks described in this book on thousands of WordPress sites. The damage can be significant and this can compromise your users' data which will become the target of attacks such as Sim Swapping, spoofing, phishing or others (see in appendix). You have the responsibility to secure your website because it usually represents a significant investment of time for you and also and above all for your users!

## Acknowledgements

Thank you for buying this book. By this gesture, you are encouraging the self-publishing movement, that of authors disseminating their works directly to their readers, without going through a publisher.

The last page offers you the possibility to rate this book in two clicks, or to write an evaluation in a few minutes. You will then be one of the 0.1% of readers who share their opinion on their reading and your opinion will make a huge difference, whether it is by encouraging other readers to discover this guide or by helping the author to improve. In either case, your comments will be greatly appreciated.

Finally, special thanks to Claudine and Nicolas who patiently reread this book to check its accuracy, readability, spelling and grammar.

# ANNEXES

## How to restore your site from backups made with BackWPup

The BackWPup plugin is very complete for the backup part, you can schedule automatic backups of your entire site (database + files) and send them to multiple locations (FTP server, Dropbox, Amazon S3, Google Storage, Microsoft Azure, RackSpaceCloud, ...). However, it does not offer any restoration functionality. This will be done manually, but don't worry, it's not complicated.

To restore the database, it will be necessary to import the .sql backup file in phpMyAdmin. It is a software accessible online from your browser, which allows you to manipulate the database of your site.

To do so, connect to phpMyAdmin using the url and the login/password combination provided by your web host. Then, in the left panel, choose the database in which you want to restore. Depending on your hosting, it is quite possible that there is only one database. If there are several databases, to be sure you have chosen the right one, check in the "Structure" tab that the database tables have the prefix you chose when you created your site (or that you have modified after reading this book), by default the prefix is "wp_", but you have already changed it.

Then, in the "Import" tab, you just have to indicate your backup file (.sql extension).

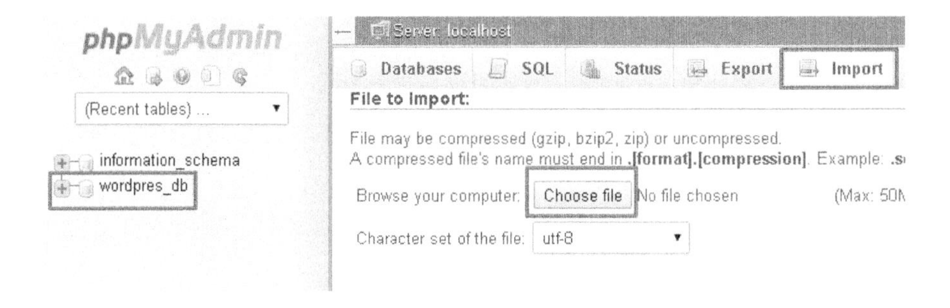


Then, you will have to uncheck the "Partial import" option, choose the "SQL" format and click on the "GO" button. The transaction will be relatively quick and you will see a message on the screen saying :

Import has been successfully finished, ## queries executed.

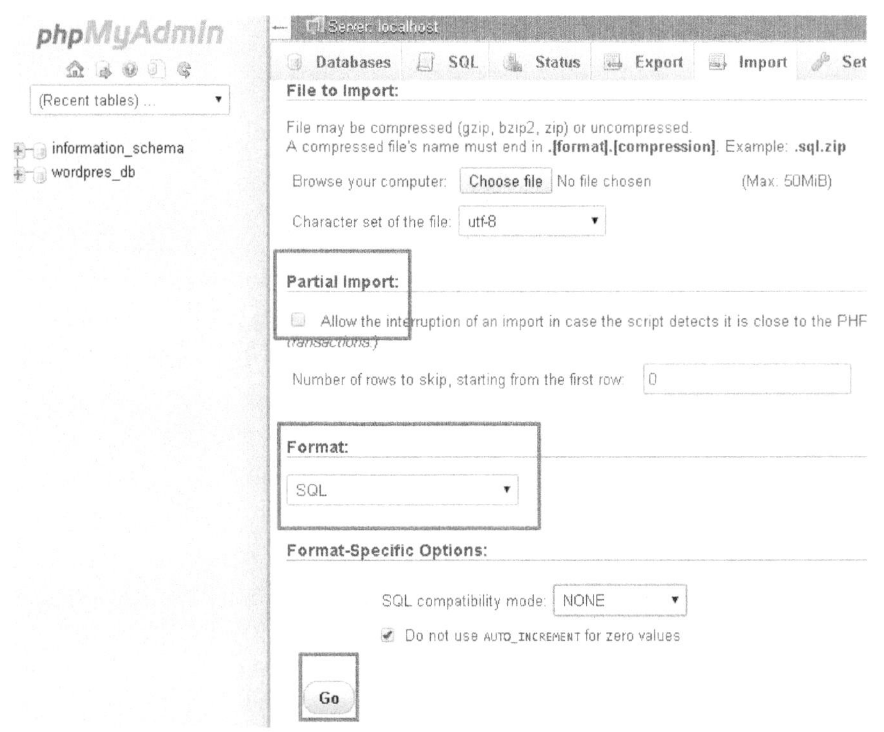


Of course, you have to make sure that the access password to phpMyAdmin is secured according to the rules recommended in this book.

If the database restoration has made your site functional, it will not be necessary to restore the files. On the contrary, if the site does not display correctly, you will have to restore the files as explained below.
To do this, connect to your hosting via FTP, using Filezilla, then delete all the files on your website and re-upload them from the backup. This operation can take a little time if your site is large, there is nothing to worry about.

# BONUS - Frequent attacks on individuals and how to protect yourself from them

## Sim swapping or SIM card duplication

The technique is simple but diabolically effective: by pretending to be you with your telephone operator, the hacker will ask for your SIM card to be reissued. Having a copy of your SIM card in his possession, the hacker will be able to impersonate you on all sites using dual authentication by SMS.

To achieve this, the hacker uses social engineering techniques to find your date of birth, your postal address, your mother's maiden name, the make of your first car... so that he can answer personal questions asked by the telephone operator to verify the identity of the caller.

But which services use SMS authentication?
Double authentication by sending a one-time use code via SMS is widely used. From "3D Secure" bank transaction validation to Amazon or Gmail account login, two-step user authentication is chosen by many services wishing to make their transactions more secure.

The direct consequence is a sharp increase in this type of attack! For example, customers of the "Coinbase" online wallet are regularly targeted by Sim Swapping.

How to protect yourself against SIM Swapping?
Rather than SMS authentication, it is clearly preferable to use two-factor authentication, using software such as authy, Google Authenticator or freeotp. Moreover, the European directive on payment services 2nd version (DSP2) recommends to stop using SMS as a means of strong authentication of Internet users.

### The usurpation of .co

Many websites are the target of a new usurpation: the purchase of the same domain name in .co instead of .com...

The system is simple: the user receives a phishing email posing as a known brand and, when logging on to the site, the domain in the address bar is not https://www.netflix.com/moncompte but https://www.netflix.co/moncompte. This fraudulent site, a mirror site reusing the HTML pages of the original site (so as to have the same appearance), will allow hackers to collect a lot of information about the victim, first of all his login and password.

These scams are extremely present on the Internet, it is absolutely necessary to check the address of the site on which one surfs before buying anything.

### Spoofing

Have you received an email from someone close to you (or even an email from yourself) that is a bit weird? Don't pay 520 euros in Bitcoin to an unknown account without thinking: you are probably a victim of spoofing. This is a method of spoofing the sending email address. This type of attack is very frequent (and sometimes credible). Usually, the hacker tries to make you believe things that are actually totally false: he has information about you, a relative needs you, etc.

### The ransomware

Other hackers go further, with ransomwares. This is an increasingly common type of attack: a person hacks a computer or server and demands a ransom, in Bitcoin, from the user or administrator. During this time, the data is blocked: the hacker forces you to pay to recover your data.

## Intrusions on connected objects

More and more people are acquiring connected objects: watches, voice assistants, lighting or security devices... There are a multitude of them for more and more widespread use in everyday life, both at home and in companies. But how vulnerable are they and what are the risks for the owners of these objects? Companies that market these solutions must redouble their efforts to build firewalls and implement recurring update programs in their design to protect them from possible attacks and prevent vulnerabilities from being exploited by malicious individuals.

## Malware on mobile phones

We now spend more time on mobile than in front of the TV, hackers have understood this and have been quick to exploit this opportunity. In 2018, 116.5 million mobile attacks occurred according to Kaspersky, almost twice as many as in 2017 (66.4 million). Some manufacturers are marketing vulnerable terminals that are prime targets for hackers and their malware, which have become more efficient and accurate. Many mobile device management solutions are available on the market, and can be used by companies to protect their employees and critical data.

## Phishing

The famous pop-up window that appears and asks you to click to retrieve the million euros you won by drawing lots, the fake email you receive from your bank asking you to enter your login and password... We have all already been confronted with phishing at least once while surfing the net. As soon as you click on a link of this type, you expose your personal data. Hackers can steal credit card numbers or financial information, as well as login credentials or confidential data by replicating input interfaces.

## Attacks against cloud storage

Many individuals and businesses have moved away from traditional storage to cloud computing that is accessible everywhere. However, hackers can use mechanisms to steal encryption keys to gain access to sensitive information and other confidential data. To counter this scourge, it is advisable to invest in a secure encryption system and an SSL certificate, provided by a trusted provider to protect your company's data.

## The use of a fake address bar on Android phones using Chrome

The latest update of Google Chrome on Android introduces a new feature initially dedicated to user comfort. Rather than displaying the address bar containing the URL of the website visited, Chrome replaces it with a simple title leaving more room to read the content itself.

But as developer James Fisher points out, it has never been easier to pass off a phishing page as a legitimate web page since the address is no longer displayed to the user. Caution is therefore recommended with this feature.

Printed in Great Britain
by Amazon